1... 2... Tree

Written by Heidi Drake
Photos by Micah & Jenny Gilday

Snow Angel Press

1...2...Tree!

Text copyright ©2012 Heidi Drake

Photography copyright © 2011 Micah & Jenny Gilday

First edition

All rights reserved. No part of this publication, other than brief excerpts for reviews, may be reproduced in any form without written permission from the publisher.

Library of Congress Control Number: 2012922077
ISBN: 978-0-9838362-0-9 (paperback)
Subjects: Counting, Hiking, Outdoors

Printed in the USA
CPSIA Section 103(a) Compliant
Batch #12/12
For further information contact:
RJ Communications, NY, NY
1-800-621-2556

Snow Angel Press
PO Box 3190
Sunriver, Oregon 97707
www.heididrakeinc.com/snowangelpress

Heidi Drake is a book-writing, dance-teaching, outdoor-loving mom who spends a lot of time hiking in the woods near her home in Sunriver, Oregon with her three kids, Elise, Maya, and Eric (the one she married).

Micah & Jenny Gilday are a photo-taking, park-exploring, life-loving pair who are lucky to be married to each other.

BIG Xs, Os and thanks to Elise, Maya, Jaida, Nolynn, Forchin, Mmynt, Harper, Conner, Sophia, their patient parents & Teacher Loekie for joining us on an unexpectedly rainy day at Eagle Fern Park, Estacada, Oregon. You guys rocked! And special thanks to Nancy, Debby, Joan, Denice, Nicole & Wendy for helping me prune 1... 2 ...Tree! so it could bloom.

To Elise & Maya, who remind me that a better attitude is just a hike away.
And to the One who created it all for us to enjoy.

H.D.

We're so honored to be part of this project. Big thanks to Heidi for having
faith in our "skills" and to all the kids who made for a
fun & inspiring day among the one…two…TREES!

M.G. & J.G.

*To Landry
From Heidi Drake*

It's a drizzly day …
Yahoo! Hurray!

No need to stay indoors.

We grab our coats and
 rain boots
  and we run out to explore!

Read this map.
What does it say?

Cross the bridge.
The fun's this way!

1 kid jumps,

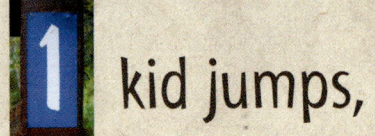

1 kid swings,

1 looks up this hollow tree.

1 kid peeks,

1 kid grins.

This one says,
"Come follow me!"

2 friends love to go for walks.

2 friends think this number rocks!

2 friends stop to share a hug,

and on the trail they see a slug!*

*That's just a snail without his shell home, you know …

3 girls stroll by.

3 boys stay dry.

3 kids hide their faces, and one's not shy.

Look: 1 stump + 2 new trunks = 3 trees in one!

4 fingers grip a leaf bouquet.

4 pairs of boots dance cold away!

4 kids goof off ~

two stick out tongues
when they hear,
"Open wide!"

Inside this big, damp,
cedar tree,

4 kids use leaves to hide.

Two kids 'high 5'…

"Hey, let's go!"

5 kids crouch low in a row.

She 'takes 5' to read
with ease
about the lives
of plants and trees.*

*Trees and plants have to die to create lush, green forests like this! It's all part of the circle of life.

This girl counts to 6. Can you?

Two thumbs up means he can too!

This mossy tree's
 the place to meet,

while three kids rest 6 muddy feet.

7 hikers play a game.
She tells him, "I'm glad you came!"

Count the letters in "BIG LEAF" (this maple's common name).

Count the laces on this boot...

The answer is the same!

You did come up with 7 both times right?

8 kids line up...
Ready, set, go!

Side by side... and in a row.

L-I-C-O-R-I-C-E has eight letters ~ this is a licorice fern!

Its name sounds like candy,
but eat this fern? No!

Its name has 8 letters …

Can you guess? Do you know?

This rubber boot has joined its friend.
Do you see 9 fake laces?

And can you spy 9 open eyes

on seven silly faces?

10 fingers cradle moist, green moss.

10 fingers give
      wet leaves a toss.

When her friends see
    what she has done …

They all jump in
and join the fun!

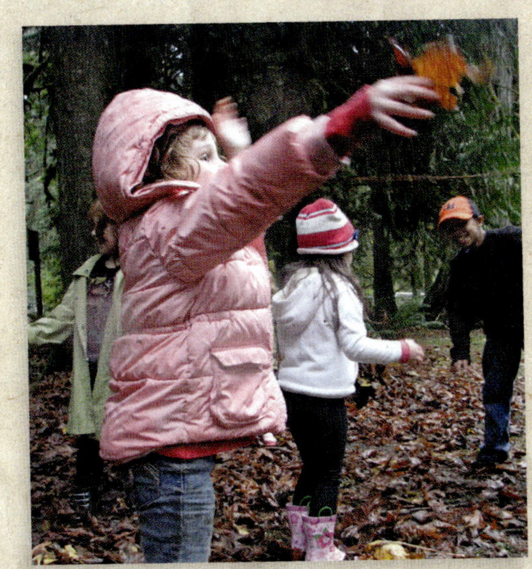

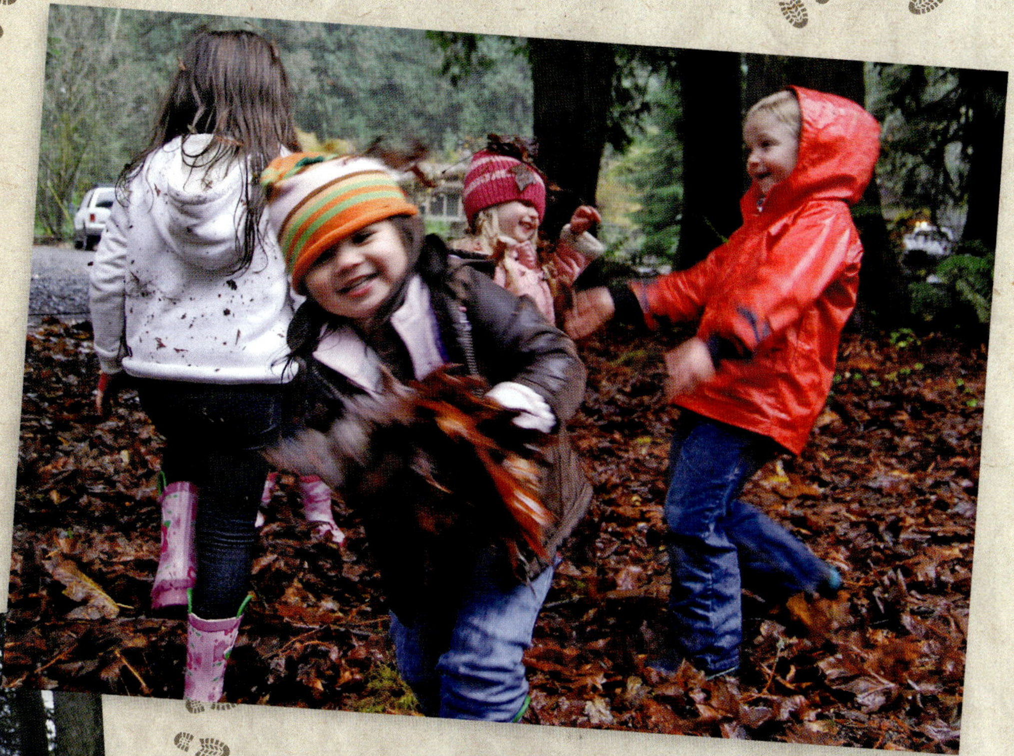

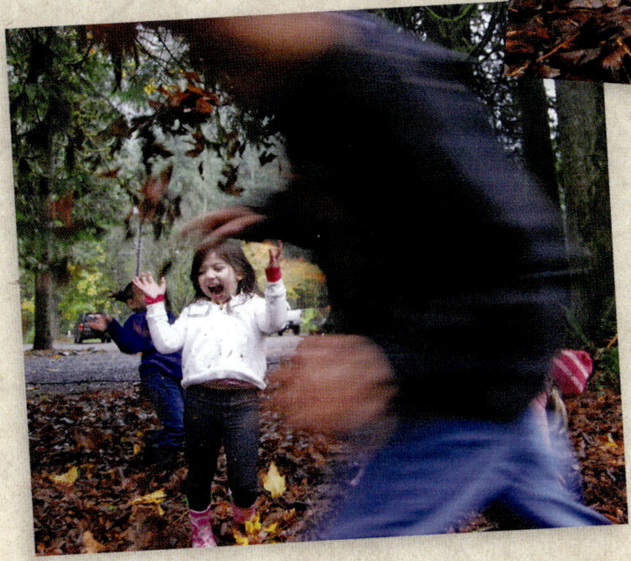

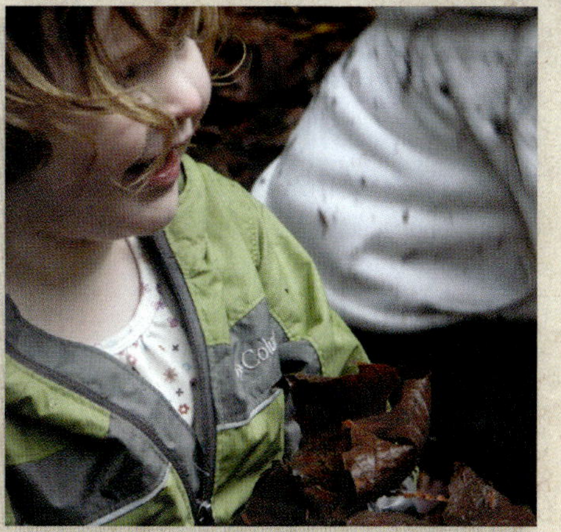

10 Hiking Tips for Outdoor Kids

1. Always hike with at least one grownup.
2. Wear hiking boots or shoes (no flip-flops!).
3. Bring a water bottle and snacks in your backpack.
4. Pack a jacket if it's cool or may rain.
5. Apply sunscreen! And bug spray if you need it.
6. Wear a whistle for safety.
7. Tell someone where you are going and when you will be back.
8. Take breaks to rest and enjoy being outside.
9. Don't let the weather stop you … it's all good.
10. Bring a camera!

10 Outdoor Books & Websites for Families

1. *First Nature Encyclopedia* by Caroline Bingham & Ben Morgan, DK Publishing
2. *Monsters in the Woods: Backpacking with Children* by Tim Hauserman
3. *Great Big Book of Children's Games: Over 450 Indoor & Outdoor Games for Kids* by Debra Wise & Sandy Forrest
4. *Nature's Playground: Activities, Crafts & Games to Encourage Children to Get Outdoors* by Fiona Danks & Jo Schofield
5. *The Kids' Nature Book: 365 Indoor/Outdoor Activities & Experiences* by Susan Milord
6. *Picture This: Fun Photography & Craft (Kids Can Do It)* by Debra Friedman
7. www.discovertheforest.org
8. www.youthoutdoorsusa.com
9. www.fws.gov/letsgooutside.com
10. www.outdoorplaces.com

10 Games You Can Play on the Trail

1. I Spy
2. Follow the Leader
3. Community Story
4. Stick Collector
5. Count the Birds
6. Track Tracker
7. Scavenger Hunt
8. Sing a Song
9. Guess What's In Your Hand
10. ABC Game

Visit www.heididrakeinc.com and click on Snow Angel Press for game instructions and tips!

Hiking Maze

Use your finger to trace a path from beginning to end.
Or have an adult make a copy of this maze so you can use a pen or pencil.

Snow Angel Press